W9-ADE-866

Millis Public Library
Auburn Road
Millis, Mass. 02054
DEC 0 8 2009

Understanding the Elements of the Periodic Table™

THE CARBON ELEMENTS

Carbon, Silicon, Germanium, Tin, Lead

Brian Belval

rosen publishing's
rosen central

New York

Published in 2010 by The Rosen Publishing Group, Inc.
29 East 21st Street, New York, NY 10010

Copyright © 2010 by The Rosen Publishing Group, Inc.

First Edition

All rights reserved. No part of this book may be reproduced in any form without permission in writing from the publisher, except by a reviewer.

Library of Congress Cataloging-in-Publication Data

Belval, Brian.
The carbon elements: carbon, silicon, germanium, tin, lead / Brian Belval.—1st ed.
     p. cm.—(Understanding the elements of the periodic table)
Includes bibliographical references and index.
ISBN-13: 978-1-4358-5334-8 (library binding)
1. Carbon—Juvenile literature. 2. Group 14 elements—Juvenile literature.
3. Periodic law—Juvenile literature. I. Title.
QD181.C1B456 2010
546'.68—dc22

                                                                2008049549

Manufactured in the United States of America

On the cover: The atomic symbols and structures of carbon, silicon, germanium, tin, and lead.

Contents

Introduction   4

Chapter One   The Carbon Group Elements   6

Chapter Two   A Closer Look at the
                  Carbon Group Elements   16

Chapter Three   Carbon Group Compounds   24

Chapter Four   The Elements in
                  Everyday Life   32

The Periodic Table of Elements   38

Glossary   40

For More Information   42

For Further Reading   44

Bibliography   45

Index   46

Introduction

The periodic table of the elements is found in chemistry textbooks, on the chemistry classroom wall, in laboratory notebooks, and all over the world in places that chemistry is practiced. Why? Because it is an extremely useful tool. The periodic table is packed with practical information that helps us understand the elements, which are the building blocks of the universe.

The more than one hundred elements known to us today are organized in the periodic table. The horizontal rows of the periodic table are known as periods, while the vertical columns are known as groups. The elements in each group share similar characteristics. Counting from left to right, the carbon group is the fourteenth column and is sometimes called group 14. There are five elements in the group—carbon at the top, and then silicon, germanium, tin, and lead. Each element is given a one- or two-letter symbol on the table. Carbon's symbol is C, silicon is Si, germanium is Ge, tin is Sn, and lead is Pb.

Tin's and lead's symbols may seem peculiar at first glance. This is because their symbols are not based on their English names. Many years ago, chemists used the Latin names of tin and lead to create their symbols. In Latin, tin is known as *stannum*, while lead is *plumbum*.

The elements of the carbon group all have numerous functions and each is important in its own unique way. Carbon is the basis for all life on our planet. All living creatures rely on it to give them structure, to provide their life processes, and for energy. Silicon is the second most common ingredient in the earth's crust after oxygen (O). Sand and rock are primarily made out of silicon. Lead and tin are both common metals. They have uses in many different industries. Germanium, although much more rare than the other elements in the carbon group, is also a useful element. It is mainly used to make parts for the computer and electronics industries.

Throughout this book, we will take a closer look at the elements of the carbon group. We will look at what they have in common, and thus why they are grouped together. We will also look at the many ways that each element is unique. Chemistry is a science that examines similarities and differences. In this book, like a good scientist, that is exactly what we will do.

Chapter One
The Carbon Group Elements

Each element in the periodic table can be described by its properties. There are two different types of properties—chemical and physical. Chemical properties have to do with how an element reacts with other elements. Some elements are very reactive; others, not so much. When elements react, they may produce heat, explode, glow, or change colors. The colorful display of a firework show is a result of chemical reactions, as is the burning of jet fuel that pushes an airplane through the air.

Chemical reactions do not always need to be explosive, colorful, or even visible. For example, in each leaf of a plant, there are thousands of chemical reactions occurring at any time. You do not notice them because they are not very energetic and are hidden within the leaf. Over time, these reactions result in noticeable changes, such as growth of the leaf or changes in the leaf's color. Chemical reactions occur around us all the time and are essential to life on our planet.

The elements of the carbon group—carbon, silicon, germanium, tin, and lead—have similar chemical properties, and that is why they are grouped together in the same column of the periodic table. For example, the elements of the carbon group all react with oxygen to form chemical compounds called oxides. A compound is made of two or more elements held together by electrical forces. These forces are known as chemical bonds.

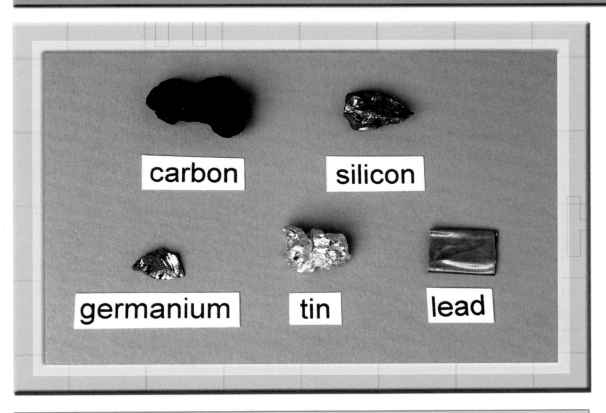

The five elements of the carbon group are solids at normal temperatures. The elements have different physical properties, such as color and density, but they all share similar chemical properties.

Physical properties have to do with an element's appearance and can be measured without reacting the element with another substance. Physical properties include color, smell, density, hardness, melting point, and boiling point. The physical properties of the elements in the carbon group vary widely. This is an important aspect of the periodic table. Although the elements in a group (column) have similar chemical properties, their physical properties may not be very similar.

One by one, let's take a closer look at the elements of the carbon group. We'll look at their physical properties in this chapter and also touch on their discovery and where they are found in nature.

Carbon

All living things contain carbon. It is also found in significant quantities in the earth's crust and in the air. Carbon was known to ancient civilizations, but they were not aware of its many functions. Ancient people were most familiar with carbon as soot and charcoal. Soot is formed from a flame that has too little air, and charcoal forms when wood is heated without air. People throughout history knew about coal, which is mostly carbon, but they did not burn it to produce energy.

Pure carbon exists in different forms—two of which are graphite and diamond. When an element has multiple forms, the various forms are known as allotropes. The graphite and diamond allotropes are very different in appearance and structure. Graphite is a black, solid substance. It consists of layers that can slide easily over each other and thus feels slippery or greasy to the touch. This property makes it a good lubricant. Graphite is found worldwide as part of the earth's crust. It is found in large quantities in South Korea, Austria, Sri Lanka, Germany, Russia, Mexico, the Czech Republic, and Italy. These countries mine graphite for use in

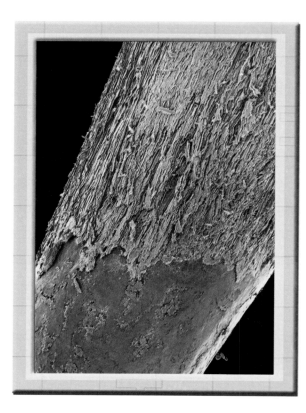

A pencil viewed under a microscope shows the wood surface on the top half and the graphite tip below it.

various industries. In addition to being useful as a lubricant, graphite is also used to make batteries, paint, and motor parts. Interestingly, the "lead" of a pencil is not lead at all, but graphite.

While graphite is black and somewhat soft and slippery, diamonds are translucent and exceptionally hard. Diamonds are, in fact, the hardest known natural substance. They are hard because the carbon in diamonds is packed in a three-dimensional network, unlike the sheets of graphite that slip and slide over each other. Another difference between graphite and diamond is the ability to conduct electricity. Graphite conducts electricity well, while diamonds do not easily allow electricity to pass through them.

Diamonds are created under great pressure deep in the earth and are found only in a few places. Mining of diamonds mostly occurs in the countries of Botswana, South Africa, Russia, Canada, Angola, and Australia. The diamond industry centers around the sale of diamonds as gemstones. Diamonds are also valuable cutting tools. Because of their extreme hardness, the point of a diamond can cut through just about any surface.

Diamond and graphite are the pure forms of carbon, but carbon also combines with other elements to form numerous compounds. Many of these compounds are found in the human body and in the food we eat. Additional compounds are found in the earth as fossil fuels, such as petroleum. Carbon compounds are exceptionally numerous and help create life as we know it.

Silicon

Silicon is the second-most common element in the earth's crust, after oxygen. Sand, rock, soil, and clay are largely composed of silicon. Silicon is typically combined with oxygen in silicon-oxygen compounds. One of

This flint knife is more than four thousand years old. Flint, like many rocks and stones, is made out of silicon.

the many natural silicon-oxygen compounds is flint, a hard material that can be easily split by hammering. Prehistoric people used flint to make sharp tools and weapons. In fact, the name "silicon" is derived from the Latin word *silicis*, meaning flint.

Silicon does not occur naturally in its pure form, and it is difficult to extract it from its natural compounds. Because of this, it took scientists a long time to realize silicon was an element. It wasn't until 1824 that the Swedish chemist Jakob Berzelius first produced pure silicon. The silicon was formed in the chemist's lab when he reacted the compound potassium fluorosilicate with potassium (K) metal.

In its pure, elemental form, silicon is a blue-gray solid with a metallic sheen. It has the same crystal structure as diamond. Although not as hard as diamond, silicon is still a very sturdy element. Its extremely high melting point of 2,570 degrees Fahrenheit (1,410 degrees Celsius) is much higher than the carbon group metals tin and lead. Silicon can conduct electricity weakly and is thus known as a semiconductor. Transistors made out of silicon are vital components of computers and many other electronic devices. The ability of miniature silicon chips to precisely control electrical current has revolutionized the computer and electronics industry.

Silicon is also an important component of solar cells. A solar cell is a device that converts sunlight to electricity. It contains a set of silicon wafers that also contain small amounts of the element arsenic (As) or boron (B). When light falls on the solar cell, electrical current flows between the silicon wafers. In the future, it is possible that solar cells will become a much more important source of energy than they are today.

Physical Properties of the Carbon Group

	State at normal room temperature	Conducts electricity?	Density (g/cm^3)	Melting point (F)	Boiling point (F)
Carbon	Solid	No (diamond) Yes (graphite)	2.27	6,422 (3,550°C)	6,917 (3,825°C)
Silicon	Solid	Partially	2.33	2,570 (1,410°C)	4,271 (2,355°C)
Germanium	Solid	Partially	5.32	1,718 (937°C)	5,126 (2,830°C)
Tin	Solid	Yes	7.29	449 (232°C)	4,118 (2,270°C)
Lead	Solid	Yes	11.34	620 (327°C)	3,164 (1,740°C)

Germanium is used to make a variety of electronic and computer products. High-quality fluorescent lamps, for example, contain small amounts of germanium.

Germanium

In 1886, a miner working deep underground in a German silver mine discovered an unusual rock. The rock was passed on to Clemens Winkler, a scientist at a nearby lab. Winkler analyzed the sample and found that it was made of mostly silver (Ag) and sulfur (S), but also an unknown substance. He was unable to break down the substance into any other substance, and so he knew that he had discovered a previously unknown element. Winkler named the element "germanium" after his native country.

Germanium is a silvery solid that looks metallic but is actually quite brittle. Germanium is not found in nature in its pure form. Instead, it is usually found mixed with other metals in the minerals germanite and argyrodite, which are rare and not mined commercially. Germanium also exists in small amounts within zinc ores. When zinc ores are processed for the zinc, germanium is isolated as a by-product. About 80 tons (72.57 metric tons) of germanium is produced yearly this way.

Germanium, like silicon, is a semiconductor. It is used in the electronics and computer industries, but less so than silicon. When chemically combined with oxygen, germanium forms a compound that is used to make high-quality camera and microscope lenses. Germanium is also used to make fluorescent lamps and infrared detectors.

Tin

Bronze is a metal made out of a mixture of tin and copper (Cu). Bronze was one of the first metals to be used by human civilization. It becomes soft when heated and was shaped into tools and weapons by master craftsmen more than five thousand years ago in Egypt and Mesopotamia. This period of history is known as the Bronze Age.

Tin is a relatively rare element but is found concentrated in a few locations in the earth's crust. The country of Malaysia produces more of the world's tin than any other country. Bolivia, Indonesia, Thailand, and Brazil also mine large amounts of tin.

Tin has two allotropes. The more common allotrope is called white tin. It is a silver-white metal that is soft and bends relatively easily. When a bar of white tin is bent, it emits a soft, high-pitched whine, called the "tin cry." The second allotrope is gray tin. It is a brittle, powdery solid. Above 56.3°F (13.5°C), tin exists as white tin. Below this temperature, white tin slowly transforms into gray tin.

Pure tin is mixed with other metals to form alloys. Bronze is an example of an alloy. Tin and lead are combined in an alloy called solder. Solder has a relatively low melting point for a metal. When heated, it turns into a liquid that is used to secure electrical connections and fasten metal parts together. Solid tin is highly malleable, meaning that it can be easily shaped, rolled, and pounded flat. This property of tin makes it a good choice for making cans and foil. Today, however, to save money, aluminum and other metals are more commonly used for these purposes.

Lead

Lead coins have been found in tombs that are more than five thousand years old, and lead pipes were used by ancient Romans to transport water in their homes. It is no coincidence that the chemical symbol for

Lead water pipes were commonly used up until the twentieth century, such as this one used to transport water during Roman times. Today, less expensive and safer materials are preferred.

lead is derived from the Latin word *plumbum*, which is also the root of the word "plumbing." Lead is a gray, highly malleable metal that can easily be worked into tools and utensils. Lead is the most dense of the elements in the carbon group. A brick-sized piece of lead weighs almost 50 pounds (23 kilograms). Lead is rarely found in its pure form in nature but instead is found mixed with other elements in ores. The most common ore of lead is galena, which is lead combined with sulfur. Australia is the number one producer of the world's lead, followed by the United States. Approximately 6 million tons (5.44 million metric tons) of lead is mined from the earth each year.

The most common use of lead is making electrodes in car and truck batteries. Small amounts of lead are used in the glass in the picture tubes of older televisions and computer monitors. The lead in the glass shields the viewer from potentially dangerous radiation generated inside the tube. This shielding property of lead also makes it useful as a part of protective vests or shields that are used when getting an X-ray. Lead is also used to make ammunition and as weights in sports equipment. As mentioned earlier, solder is an alloy of lead and tin with many uses in manufacturing.

Lead is toxic to the human body. When ingested in high enough amounts or over a long period of time, it can cause lead poisoning. Lead was used in making paint because it produced a white pigment that covered old paint well. However, because lead is toxic, dust from this old paint is now causing serious problems for people who live in old houses where lead paint was used. Lead was also used in making glaze for pottery because it produced a glaze that melted at low temperature, making the pottery easy to glaze. However, when food was stored in lead-glazed pottery, some of the lead got into the food and poisoned the people who ate the food. Because of this, the use of lead in paint and pottery and other products has been greatly reduced in the last few decades.

Chapter Two
A Closer Look at the Carbon Group Elements

An element is a substance that cannot be broken down into other substances. If you shock an element with electricity, blast it with dynamite, heat it at a high temperature, add strong acid to it, or anything else, it will not break down into something else. There are ninety-two naturally occurring elements. An additional twenty or so have been created in laboratories. These man-made elements do not exist in nature and are so unstable that most of them exist for a matter of seconds after creation. All of the elements are listed in the periodic table.

Each element is made up of tiny particles called atoms. All of the atoms of an element are alike. Different elements have different atoms. The atoms differ in size and weight.

The element with the lightest atoms, hydrogen (H), has existed since the formation of the universe billions of years ago. The big bang also created the elements helium (He) and lithium (Li). If you look at the periodic table, you will see that hydrogen, helium, and lithium are the first three elements listed.

The rest of the elements were created many years after the big bang inside of stars. Inside stars, gravity produces an enormous pressure. This pressure squeezes the atoms of hydrogen, helium, and lithium together so hard that they merge together into bigger atoms. When atoms merge together, they release an enormous amount of energy.

The elements are created by stars like the giant nebula NGC 3603. Energy inside a star can combine smaller elements into bigger ones. Elements are also formed by the energy created when stars explode.

Our sun is a star, and the energy we get from it comes from merging smaller atoms into larger ones. Carbon and silicon in the carbon group were created this way. Eventually, the stars that created the elements burned out. As they burned out, they exploded, and the tremendous energy released caused atoms to merge into even larger atoms, producing germanium, tin, and lead. The explosion also scattered all the elements into space. Over time, gravity pulled on the elements, and they were shaped into planets. Earth and the other planets became enveloped in an atmosphere of light gases, while the core and crust were made out of mostly heavier elements. Humans have inhabited our planet for only a tiny fraction of the history of the universe. The elements

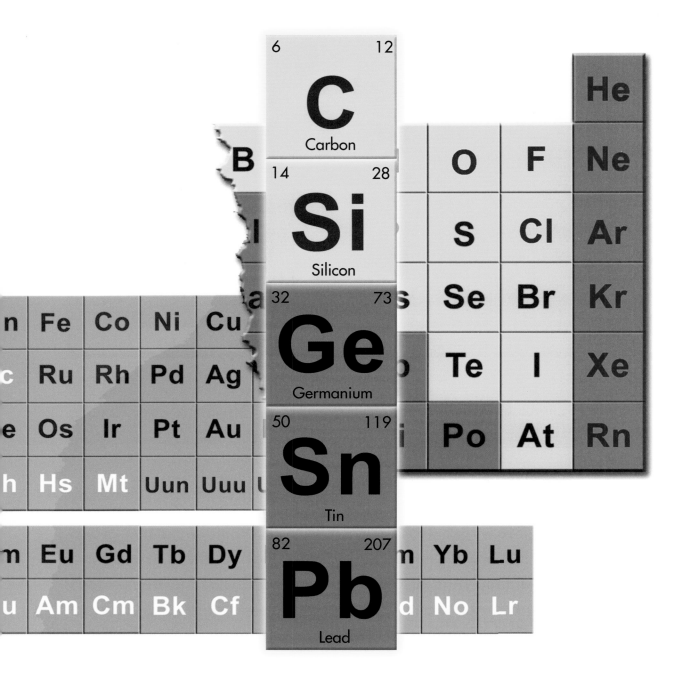

The elements of the carbon group appear in the 14th column of the periodic table. The element in the group with the lowest atomic number (carbon) is at the top. The element with the highest atomic number (lead) is at the bottom.

are our building blocks, making up our bodies, the air we breathe, and the ground we walk on.

Atoms

An atom is the smallest particle of an element. Atoms themselves are made out of even smaller particles—electrons, protons, and neutrons. Neutrons and protons form the center, or nucleus, of the atom. The nucleus contains most of the weight of an atom. Protons have a positive electrical charge, and neutrons have no electrical charge. Every atom of an element has the same number of protons. For example, all carbon

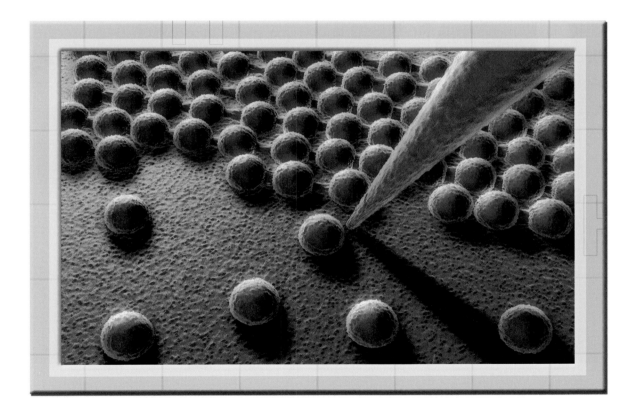

Individual atoms are visible when viewed by a high-powered microscope. The four atoms at the bottom of the photo were once bonded to the cluster of atoms in the top half of the photo.

atoms have six protons, silicon atoms have fourteen protons, germanium atoms have thirty-two protons, tin atoms have fifty protons, and lead atoms have eighty-two protons. The number of protons defines the element. The number of protons is known as the atomic number. If you look at the periodic table, you will see the atomic number listed, usually above the element's symbol. The atomic number increases as you move left to right on the table, or down the columns of the table.

Electrons are found outside the nucleus of the atom. They have a negative electrical charge. Electrons are much smaller than neutrons and protons but are very energetic. They dart and flutter around the nucleus. Because of the pull that the positively charged nucleus has on the negatively charged electrons, the electrons are limited in how far they can stray. They move in limited paths known as shells or orbitals. Because the electric charge needs to be balanced, the number of electrons in an atom is equal to the number of protons. Sometimes, an atom can gain or

The Carbon Group Elements and the Periodic Table

	Symbol	Atomic number (number of protons)	Atomic weight	Number of electrons in outer shell
Carbon	C	6	12.01	4
Silicon	Si	14	28.09	4
Germanium	Ge	32	72.59	4
Tin	Sn	50	118.7	4
Lead	Pb	82	207.2	4

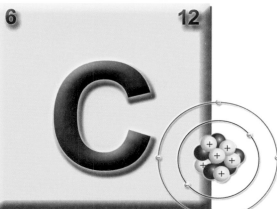

The nucleus and electron shells of each element of the carbon group are pictured. Within the nucleus, the spheres with a + sign are protons, while the darker spheres are neutrons.

lose an electron to become negatively or positively charged. When this happens, the atom is called an ion.

The Role of Electrons

In each atom, the electrons are arranged in shells. Carbon has two shells, silicon has three, germanium has four, tin has five, and lead has six. If you look at the periodic table, you will see that the number of electron shells is equal to the number of the row in which the element resides. This is true of all the elements in the table. The row number equals the number of electrons shells.

An atom is most stable when its outer shell is full of electrons. These outer shell electrons are called valence electrons. Most of the time, a full outer shell consists of eight electrons. Atoms will seek out electrons or give up electrons so that their outer shell is full. Much of

Element 114

Beneath lead in many versions of the periodic table is an element with the symbol Uuq. This element is named ununquadium, which means 114 in Latin. This element is man-made. It was first created by Russian scientists in 1998, when they crashed atoms of plutonium (Pu) and calcium (Ca) into each other. The collision resulted in the formation of only a single atom of ununquadium. The ununquadium was very unstable and broke down after about thirty seconds. A few more atoms of ununquadium have been created since then, but those atoms also broke down after a short time.

their chemical behavior is driven by this need to have a full outer shell of electrons.

All of the elements in the carbon group have four valence electrons. These four electrons interact with the outer electrons of other atoms they encounter. Sometimes, two atoms will share or trade electrons. This is known as chemical bonding. A classic example of a chemical bond can be illustrated with carbon and hydrogen. Carbon has four valence electrons, while the hydrogen atom has only one. In order to become more stable, a carbon atom will form bonds with four hydrogen atoms. Now, it has its original four electrons, plus the four electrons offered by hydrogen, for a total of eight in its outer shell. The chemical formula for the newly formed compound is CH_4, also known as methane.

Chemical Bonds

All of the elements in the carbon group form bonds in a similar way. Because of this, they are said to have similar chemical properties. In the chemical world, there are two types of bonds—covalent and ionic. In a covalent bond, the electrons are shared by two atoms. The bonds in the methane molecule are examples of covalent bonds. Ionic bonds occur when electrons are transferred from one atom to another. In such situations, the atom that receives the electrons becomes a negatively charged ion. The atom that gives up the electrons becomes a positively charged ion. The two ions are held together by the electrical attraction between the positive ion and the negative ion.

Chapter Three
Carbon Group Compounds

A compound is a substance made up of two or more elements bound together by chemical bonds. The carbon group elements form a vast number of different compounds. These compounds can be found in our bodies, in the food we eat, deep inside the earth, and even in the dark reaches of distant galaxies.

Carbon Compounds

Carbon is not the most abundant element on the earth, but it is certainly the most important. All living organisms depend on carbon for energy. It is also the building block of the structures that make all plant and animal life possible. A tree trunk is made out of carbon compounds, as are a plant's leaves. So are the muscles, bones, organs, and all the cells of the human body. Carbon compounds are so varied and complex that an entire branch of chemistry, called organic chemistry, is dedicated to studying them.

Carbon's wonderful versatility is due to the arrangement of its outer shell electrons. The four valence electrons allow carbon to form bonds to four other atoms. Carbon atoms easily form bonds to other carbon atoms. This allows carbon to form very complex molecules containing many carbon atoms. These complex molecules can be either very strong or very reactive. Because of this, carbon compounds can react and change quickly from one carbon compound to another. This constant movement of electrons between carbon atoms and other atoms is essential for life as we know it.

The Carbon Cycle

Carbon is so essential that nature has devised a system to recycle it. The carbon cycle is the exchange of carbon between living things, the atmosphere, the oceans, and the earth's crust. Carbon is never destroyed during the carbon cycle; it only changes form.

One of the most important carbon compounds in the cycle is carbon dioxide. This compound is made up of one atom of carbon combined with two atoms of oxygen. It has the chemical formula CO_2. Carbon dioxide in the atmosphere is absorbed by plants and, with the help of energy from the sun, is converted into a carbon compound called glucose ($C_6H_{12}O_6$). This process of converting carbon dioxide into glucose is known as photosynthesis. In this process, the energy of the sun is stored in the molecules of glucose. The plant uses this energy to grow. Eventually, either the plant will be eaten or it will die. If it dies, it decomposes. As it decomposes, bacteria and other organisms extract energy from the glucose in the plant by converting it back into carbon dioxide, which they release into the atmosphere. If the plant is eaten, the glucose is converted into energy and carbon dioxide by the animal or human that eats it. The carbon dioxide is released back into the atmosphere when the animal or human breathes. This process is called respiration. The cycle begins again when the carbon dioxide in the atmosphere is taken in by another plant or tree.

Hydrocarbons

Hydrocarbons, as the name suggests, are compounds made solely out of hydrogen and carbon. Hydrocarbons are the primary type of fuel

worldwide. Methane (CH_4) is the most simple hydrocarbon. Methane is also known as natural gas. Butane (C_4H_{10}) is the fuel inside a pocket lighter. Oil and gasoline are made out of various hydrocarbons, such as heptane (C_7H_{16}) and octane (C_8H_{18}). When hydrocarbons are burned as fuel, carbon dioxide is released into the atmosphere. Water is another product of this reaction, which is called combustion.

Carbon Compounds in the Human Body

The human body requires numerous carbon compounds. Proteins are required for healthy muscles. Carbohydrates and sugars, such as glucose, are the primary source of energy, while lipids (a type of fat) are required for healthy, functioning cells. DNA, the genetic material inside each of your cells, is also a carbon compound.

Silicon Compounds

Like carbon, and all the carbon group elements, silicon has four electrons available for bonding. Also like carbon, silicon often bonds with oxygen. Silicon dioxide (SiO_2), also known as silica, is a common silicon-oxygen compound. The mineral quartz is made of silicon dioxide. The atoms in silicon dioxide in quartz are arranged in a tightly packed crystal structure, much like the carbon atoms in a diamond.

The silicates are a group of silicon-oxygen compounds found in most rocks, soils, and clays. The silicates are similar to silica, except that they have more than two oxygen atoms per silicon atom. There are always more electrons than protons in a silicate, and therefore silicates

In the atmosphere and oceans, carbon exists mostly as carbon dioxide, which is converted by plants into sugar. The carbon-containing sugar is then used as a source of energy by animals.

Amethyst is a type of quartz. In addition to being very attractive, amethyst is also extremely hard and durable. These properties make it a popular gem in rings and other pieces of jewelry.

always contain negatively charged silicate ions. In rocks and minerals, the negatively charged silicate ions combine with a positively charged metal. For example, in the mineral enstatite, a positively charged magnesium (Mg) ion bonds with a negatively charged silicate ion (SiO_3) to form $MgSiO_3$. In nature's rocks, there are hundreds of different combinations of various metals and silicate ions.

Silicones are man-made compounds made out of silicon. They consist of long chains of alternating silicon and oxygen atoms. Other elements, such as carbon and hydrogen, often branch off from the silicon atoms. Silicone is used to make rubber, lubricants, sealants, hose and tubing, hair conditioner, and many different kinds of toys. Silicone was invented by the British chemist Frederic Kipping in the 1920s. However, the process that he used to make the material was very difficult to duplicate. The American chemist Eugene Rochow revolutionized silicon production in the 1940s. He found that he could mass-produce silcone by heating elemental silicon at high temperature and pressure and then adding a chlorine compound. Today, more than 500,000 tons (453,592 metric tons) of silicone are manufactured each year.

Dmitry Mendeleyev

In 1869, a Russian chemistry professor named Dmitry Mendeleyev (1834–1907) created the first periodic table. His table consisted of rows and columns and looked similar to today's periodic table, except it contained only sixty-two elements. Interestingly, Mendeleyev left gaps in his table. He did this because he suspected that there were numerous elements yet to be discovered. Over time, Mendeleyev's genius was proven when a number of new elements were discovered and fit perfectly in the gaps.

Germanium, Tin, and Lead Compounds

Germanium combines with oxygen to form germanium dioxide (GeO_2). It forms naturally when pure germanium comes into contact with air. Germanium dioxide is used to make high-quality camera and microscope lenses. Germane (GeH_4) is similar in structure to methane (CH_4). This flammable gas is used by the semiconductor industry to make transistors and other computer and electronics parts. Argyrodite (Ag_8GeS_6), or silver germanium sulfide, is the most common mineral that contains germanium. It is so rare, however, that it is rarely mined for the germanium inside it.

Tin also combines with oxygen to form tin oxide (SnO_2). The mineral cassiterite is a natural form of tin oxide. It is commonly mined for its valuable tin content. Tin reacts with chlorine to form tin chloride ($SnCl_2$). This white solid can be dissolved in water and is used to coat steel with a thin layer of tin. This type of steel, known as tinplate, is used to make food containers. The thin layer of tin keeps the food from reacting with the steel. Tinplate, however, is not as popular as it once was in food packaging. Today, containers are often made from a nonreactive steel without the tin layer, or a less expensive

Tin fluoride is a common ingredient in toothpaste. The compound helps to destroy bacteria in the mouth and prevents tooth decay.

type of metal is used to plate the steel. Another interesting tin compound is tin fluoride (SnF_2). Also known as stannous fluoride, this compound is a common ingredient in toothpaste. Fluoride, as you know, helps prevent tooth decay. It cannot be added to the paste on its own, however, so it is bonded to tin in a nontoxic compound.

Lead, like all the carbon group elements mentioned here, forms a compound with oxygen. Lead oxide (PbO_2) is used primarily in the manufacture of car batteries. It is also used as a glaze for ceramic plates and bowls. Lead is found in nature as the mineral galena, which is a lead-sulfur compound (PbS). Galena is easily mined and refined into pure lead. Lead compounds have traditionally been used to make various colors of paint. For example, trilead tetraoxide (Pb_3O_4) is used to make reddish brown paint, lead carbonate ($PbCO_3$) to make white paint, and lead chromate ($PbCrO_4$) to make yellow paint. However, we now know that lead becomes poisonous as it is absorbed over time by the body. Because of this, the use of lead-based paint in the United States has been banned since 1978 in schools, public buildings, and residential homes.

Chapter Four
The Elements in Everyday Life

The carbon group elements are incredibly important to us. The elements have valuable uses in medicine, the arts, and numerous industries, including electronics, computers, automotive, and manufacturing. We simply cannot live without some of the elements, especially carbon. Here are some of the places carbon elements can be found.

Sugar

Who doesn't have a sweet tooth? Sugar is probably the world's favorite compound made out of carbon, hydrogen, and oxygen. These sticky, sweet carbohydrates are a multibillion-dollar industry. The three most common types of sugar are sucrose, fructose, and lactose. Sucrose is also known as table sugar. It is found in sugar cane and sugar beet. Fructose ($C_6H_{12}O_6$) is found in fruits and some vegetables. Lactose is found in milk.

Carbon Dating

Have you ever wondered how they determine the age of ancient artifacts? Carbon dating is the answer. Carbon-14 is a radioactive isotope of carbon that contains eight neutrons instead of the usual six. Every living thing has a little bit of carbon-14 in it. The extra neutrons make

Much of the sugar consumed in the United States is refined sugar, also known as table sugar or sucrose. It comes from two plants—sugar beet and sugarcane.

carbon-14 atoms unstable, and the atoms begin to decay over time. By comparing the amount of carbon-14 in a sample to the amount of carbon-12 (carbon atoms with six neutrons), scientists can figure out the age of a sample.

Glass

Did you know that glass is made out of sand? Silica (SiO_2), which is what most sand is made out of, is the main ingredient in glass. Glass is made by mixing silica with soda ash (sodium carbonate) and limestone (calcium carbonate). Various other compounds can be added to this basic mix to add color or to alter the properties of the glass. The mixture is then melted and blown or molded into the desired shape.

Carbon Monoxide

Carbon monoxide (CO) is a compound whose molecules contain a single atom of carbon bonded to a single atom of oxygen. It is a colorless, odorless, and tasteless gas. It is also highly poisonous. Usually, carbon dioxide is formed when carbon compounds burn. But when oxygen levels are low, carbon monoxide can form. Older or poorly working stoves can produce dangerous levels of carbon monoxide in the home. Inhaling carbon

Silica, the primary ingredient in glass, has a very high melting point. Adding other ingredients to silica, such as sodium carbonate, lowers the melting point and makes it easier for glassblowers to work with.

monoxide produces severe headaches, nausea, dizziness, and eventually, death. According to the U.S. Environmental Protection Agency, hundreds of people die each year due to carbon monoxide poisoning. They recommend that all fuel-burning appliances be properly maintained and that you make sure you have a carbon monoxide detector in your home.

Talc

Talc is a mineral with a complex chemical formula of $Mg_3Si_4O_{10}(OH)_2$. It is a silicate that is resistant to heat, electricity, and acids. It is also the primary ingredient in talcum powder and baby powder. Talc helps absorb moisture

Carbon and Climate Change

A large amount of carbon is stored in fossil fuels. The fossil fuels—oil, coal, and natural gas—were formed from the remains of plants and animals many millions of years ago. For much of the earth's history, the fossil fuels have rested underground, untouched by man. In the last hundred years, however, humans have begun to dig up large amounts of fossil fuels. When these fuels are burned, carbon dioxide is released into the atmosphere, along with many pollutants. So much carbon dioxide is being released that it is disturbing the balance of the carbon cycle. When there is too much carbon dioxide in the air, heat becomes trapped near the earth. As a result, the planet's surface temperature has increased noticeably in the past decades. This phenomenon is often referred to as global warming.

when sprinkled on the skin. This helps reduce odor and prevents rashes and bacterial growth.

Ceramics

For dinner, you probably eat off of a plate made from silicon. Your soup bowl probably contains silicon, too. Ceramics are made from clay, which is a silicate. The structure of clay allows it to absorb large amounts of water. Wet clay is easily shaped and then can be placed in a hot furnace to harden. Porcelain, earthenware, stoneware, and fine china are all different ceramics.

Concrete

The sidewalk in front of your house or apartment is made out of silicon compounds and assorted other compounds. Concrete is a mixture of cement,

An artisan shapes a vase using a potter's wheel. The element silicon is found in all clay pottery.

gravel, sand, and water. There are different types of cement, but Portland cement is most commonly used in making concrete. Portland cement is a paste made out of silicon oxide, calcium oxide, aluminum oxide, ferric oxide, and magnesium oxide. Beyond sidewalks, concrete is used to build skyscrapers, bridges, superhighways, houses, and dams.

Tin in Your Teeth

Your dentist has a special use for tin. A mixture of silver, tin, mercury (Hg), and copper is used to fill cavities in teeth. The mixture is known as an amalgam. The amalgam starts as a soft paste that quickly hardens and expands to create a durable filling. Amalgam fillings have become less popular recently because of concerns that mercury in the filling may be harmful to the body. However, the U.S. Food and Drug Administration and American Dental Association have stated that they believe the fillings are safe to use.

Lead Acetate

Hopefully, you have no need to use this compound yet. Lead acetate is an ingredient in some hair dyes that turn gray hair dark. Lead acetate products

Tin is often a component of dental fillings. Regular brushing is the best way to avoid cavities and the need for these expensive treatments.

need to be applied over a long time to have any effect. People who use lead acetate products have been studied to make sure that the lead doesn't get absorbed into the bloodstream. If it did, it could result in lead poisoning. Thankfully, there is no evidence that lead acetate in hair dyes gets into the bloodstream.

Carbon Group Elements and Tomorrow

Tomorrow's scientists, artists, and engineers will surely find new uses for the carbon group elements. These versatile elements have endless possibilities limited only by our imaginations. Of course, with every new technology comes potential benefit and also potential harm. The burning of fossil fuels for energy, for example, has made it easier to get from point A to point B, but it has also polluted our atmosphere with dangerous chemicals. A better understanding of chemistry and the other sciences can help society improve the environment and also help us to appreciate it more. Nature has loaned us the elements to take care of during our lifetimes. It is our duty to take good care of them and leave the world a better place than ever for our ancestors.

The Periodic Table of Elements

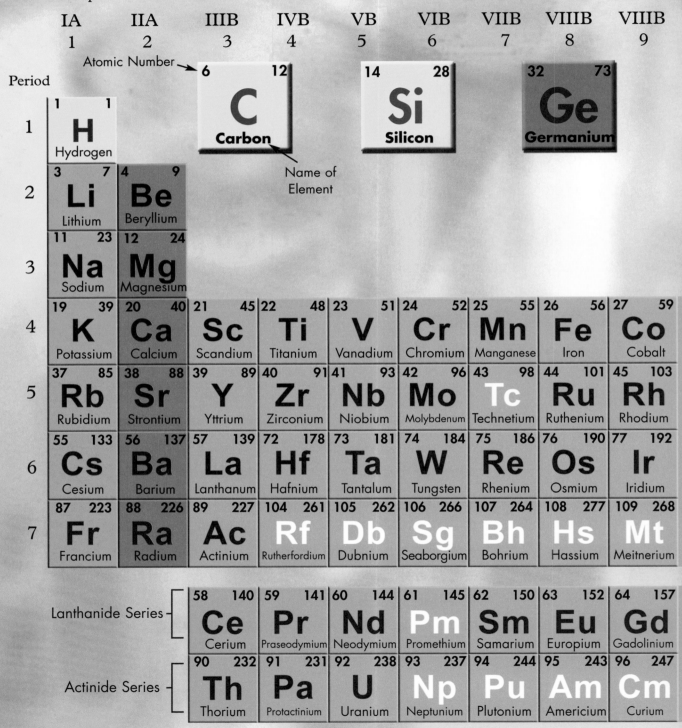

Group

IA	IIA	IIIB	IVB	VB	VIB	VIIB	VIIIB	VIIIB
1	2	3	4	5	6	7	8	9

Atomic Number →

6 **12**
C
Carbon

14 **28**
Si
Silicon

32 **73**
Ge
Germanium

Name of Element

Period

1

1 1
H
Hydrogen

2

3 7
Li
Lithium

4 9
Be
Beryllium

3

11 23
Na
Sodium

12 24
Mg
Magnesium

4

19 39
K
Potassium

20 40
Ca
Calcium

21 45
Sc
Scandium

22 48
Ti
Titanium

23 51
V
Vanadium

24 52
Cr
Chromium

25 55
Mn
Manganese

26 56
Fe
Iron

27 59
Co
Cobalt

5

37 85
Rb
Rubidium

38 88
Sr
Strontium

39 89
Y
Yttrium

40 91
Zr
Zirconium

41 93
Nb
Niobium

42 96
Mo
Molybdenum

43 98
Tc
Technetium

44 101
Ru
Ruthenium

45 103
Rh
Rhodium

6

55 133
Cs
Cesium

56 137
Ba
Barium

57 139
La
Lanthanum

72 178
Hf
Hafnium

73 181
Ta
Tantalum

74 184
W
Tungsten

75 186
Re
Rhenium

76 190
Os
Osmium

77 192
Ir
Iridium

7

87 223
Fr
Francium

88 226
Ra
Radium

89 227
Ac
Actinium

104 261
Rf
Rutherfordium

105 262
Db
Dubnium

106 266
Sg
Seaborgium

107 264
Bh
Bohrium

108 277
Hs
Hassium

109 268
Mt
Meitnerium

Lanthanide Series

58 140
Ce
Cerium

59 141
Pr
Praseodymium

60 144
Nd
Neodymium

61 145
Pm
Promethium

62 150
Sm
Samarium

63 152
Eu
Europium

64 157
Gd
Gadolinium

Actinide Series

90 232
Th
Thorium

91 231
Pa
Protactinium

92 238
U
Uranium

93 237
Np
Neptunium

94 244
Pu
Plutonium

95 243
Am
Americium

96 247
Cm
Curium

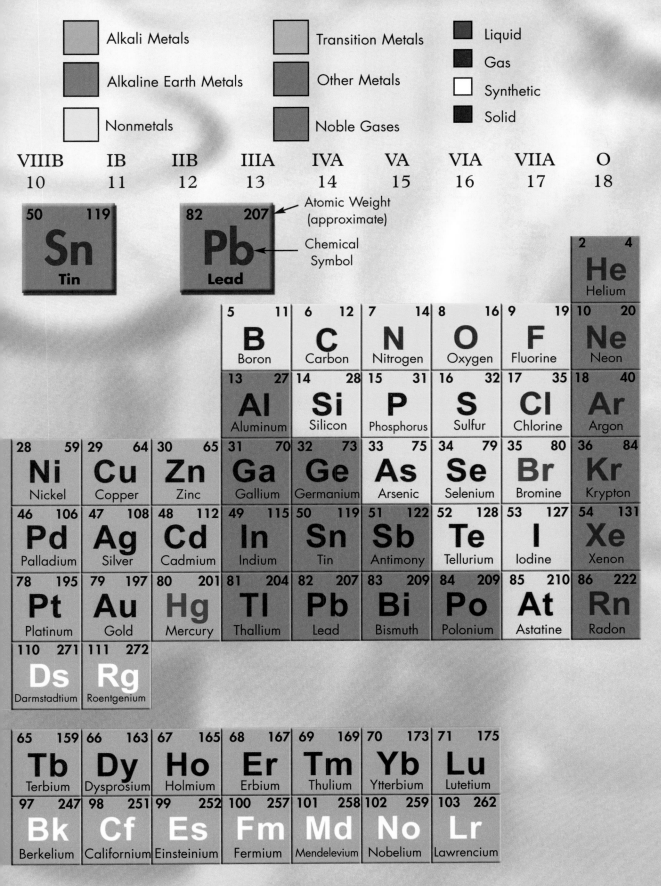

Alkali Metals

Alkaline Earth Metals

Nonmetals

Transition Metals

Other Metals

Noble Gases

Liquid

Gas

Synthetic

Solid

VIIIB	IB	IIB	IIIA	IVA	VA	VIA	VIIA	O
10	11	12	13	14	15	16	17	18

50	119
Sn	
Tin	

82	207
Pb	
Lead	

Atomic Weight (approximate)

Chemical Symbol

2	4
He	
Helium	

5 11	6 12	7 14	8 16	9 19	10 20
B Boron	**C** Carbon	**N** Nitrogen	**O** Oxygen	**F** Fluorine	**Ne** Neon
13 27	14 28	15 31	16 32	17 35	18 40
Al Aluminum	**Si** Silicon	**P** Phosphorus	**S** Sulfur	**Cl** Chlorine	**Ar** Argon

28 59	29 64	30 65	31 70	32 73	33 75	34 79	35 80	36 84
Ni Nickel	**Cu** Copper	**Zn** Zinc	**Ga** Gallium	**Ge** Germanium	**As** Arsenic	**Se** Selenium	**Br** Bromine	**Kr** Krypton
46 106	47 108	48 112	49 115	50 119	51 122	52 128	53 127	54 131
Pd Palladium	**Ag** Silver	**Cd** Cadmium	**In** Indium	**Sn** Tin	**Sb** Antimony	**Te** Tellurium	**I** Iodine	**Xe** Xenon
78 195	79 197	80 201	81 204	82 207	83 209	84 209	85 210	86 222
Pt Platinum	**Au** Gold	**Hg** Mercury	**Tl** Thallium	**Pb** Lead	**Bi** Bismuth	**Po** Polonium	**At** Astatine	**Rn** Radon
110 271	111 272							
Ds Darmstadtium	**Rg** Roentgenium							

65 159	66 163	67 165	68 167	69 169	70 173	71 175
Tb Terbium	**Dy** Dysprosium	**Ho** Holmium	**Er** Erbium	**Tm** Thulium	**Yb** Ytterbium	**Lu** Lutetium
97 247	98 251	99 252	100 257	101 258	102 259	103 262
Bk Berkelium	**Cf** Californium	**Es** Einsteinium	**Fm** Fermium	**Md** Mendelevium	**No** Nobelium	**Lr** Lawrencium

Glossary

alloy A mixture of two or more metals.

atmosphere The air surrounding the earth.

atom The smallest unit of an element. The atom itself is made out of electrons, protons, and neutrons.

combustion Chemical reaction between a substance and oxygen that gives off light and heat. Also known as burning.

compound A substance made up of two or more elements bound together by chemical bonds. The elements in a compound combine in fixed ratios, such as two atoms to one or three atoms to two.

conduct To allow something, such as heat or electricity, to pass through.

crystal A solid in which the components are arranged in a highly ordered, repeating pattern.

current The flow of electrically charged particles.

density The mass of a sample divided by its volume.

electrode An electrical substance or wire attached to any electronic device that causes current to flow.

electron A negatively charged subatomic particle found outside of the nucleus, or center, of an atom.

infrared A type of light with waves longer than red light.

ion A positively or negatively charged atom.

malleable Capable of being bent or hammered into a shape.

mineral A naturally occurring element or compound, typically found at the surface or beneath the surface of the earth.

molecule Two or more atoms joined together by chemical bonds. For example, a water molecule consists of two atoms of hydrogen and one atom of oxygen chemically attached to each other. A molecule is the smallest unit of a substance that still retains its chemical properties.

neutron A subatomic particle without charge that is found in the nucleus of an atom.

ore Rock that contains a valuable metal.

oxide A compound made of oxygen and another element.

proton A positively charged subatomic particle that is found in the nucleus of an atom.

radiation Energy given off by unstable atoms.

semiconductor A substance that conducts only a slight electrical current.

silicate A type of rock-forming mineral that contains silicon and oxygen.

soot Tiny, black particles of carbon that result from the incomplete burning of carbon-containing substances.

transistor A device that controls the flow of electricity.

valence electron An electron in the outermost shell of an atom.

For More Information

American Chemical Society
1155 Sixteenth Street NW
Washington, DC 20036
(800) 227-5558
Web site: http://www.acs.org
The American Chemical Society is the world's largest scientific society. It
  is very active in chemistry education worldwide.

Center for Science and Engineering Education
Lawrence Berkeley National Laboratory
1 Cyclotron Road MS 7R0222
Berkeley, CA 94720
(510) 486-5511
Web site: http://csee.lbl.gov
The CSEE is an organization dedicated to training the next generation
  of scientists and engineers.

International Union of Pure and Applied Chemistry
IUPAC Secretariat
P.O. Box 13757
Research Triangle Park, NC 27709-3757
(919) 485-8701
Web site: http://www.iupac.org/general/FAQs/elements.html
The IUPAC is the official organization behind the periodic table. It
  approves new elements and makes sure all of the information in the
  table is accurate and up to date.

Jefferson Lab
12000 Jefferson Avenue
Newport News, VA 23606
(757) 269-7100
Web site: http://education.jlab.org
Jefferson Lab is operated by the U.S. Department of Energy. Scientists at
    Jefferson primarily study the nucleus of the atom.

Los Alamos National Laboratory
P.O. Box 1663
Los Alamos, NM 87545
(888) 841-8256
Web site: http://periodic.lanl.gov
Scientists at Los Alamos work in the fields of chemistry, computer sci-
    ence, earth and environmental sciences, materials science, and
    physics. Many of the world's top scientists work at Los Alamos.

Web Sites

Due to the changing nature of Internet links, Rosen Publishing has developed
an online list of Web sites related to the subject of this book. This site is
updated regularly. Please use this link to access the list:

http://www.rosenlinks.com/uept/tce

For Further Reading

Baldwin, Carol. *Metals*. Chicago, IL: Raintree, 2004.

Blobaum, Cindy. *Periodic Table: Critical Thinking and Chemistry*. Austin, TX: Prufrock Press, 2005.

David, Laurie, and Cambria Gordon. *Down-to-Earth-Guide to Global Warming*. New York, NY: Scholastic, 2007

Desonie, Dana. *Atmosphere: Air Pollution and Its Effects*. New York, NY: Chelsea House, 2007.

Ham, Becky. *The Periodic Table*. New York, NY: Chelsea House, 2008.

Levy, Janey. *Tin*. New York, NY: Rosen Publishing, 2009.

Lew, Kristi. *Lead*. New York, NY: Rosen Publishing, 2008.

Manning, Philip. *Atoms, Molecules, and Compounds*. New York, NY: Chelsea House, 2007.

Saucerman, Linda. *Carbon*. New York, NY: Rosen Publishing, 2004.

Saunders, Nigel. *Carbon and the Elements of Group 14*. Chicago, IL: Heinemann, 2003.

Sommers, Michael. *Silicon*. New York, NY: Rosen Publishing, 2007.

Stille, Darlene. *Atoms & Molecules: Building Blocks of the Universe*. Mankato, MN: Capstone, 2007.

Yarrow, Joanna. *How to Reduce Your Carbon Footprint*. San Francisco, CA: Chronicle Books, 2008.

Bibliography

Emsley, John. *Nature's Building Blocks: An A–Z Guide to the Elements*. Oxford, England: Oxford University Press, 2001.

NASA Earth Observatory. "The Carbon Cycle." Retrieved September 20, 2008 (http://earthobservatory.nasa.gov/Library/CarbonCycle).

Portland Cement Association. "Concrete Basics." Retrieved September 20, 2008 (http://www.cement.org/basics/concretebasics_concretebasics.asp).

Roston, Eric. *The Carbon Age: How Life's Core Element Has Become Civilization's Greatest Threat*. New York, NY: Walker, 2008.

Royal Society of Chemistry. "A Visual Interpretation of the Table of Elements." Retrieved September 20, 2008 (http://www.rsc.org/chemsoc/visualelements).

Strathern, Paul. *Mendeleyev's Dream: The Quest for the Elements*. New York, NY: St. Martin's Press, 2000.

Stwertka, Albert. *A Guide to the Elements*. New York, NY: Oxford University Press, 2002.

Thomas Jefferson National Accelerator Facility. "It's Elemental: The Periodic Table of Elements." Retrieved September 20, 2008 (http://education.jlab.org/itselemental/index.html).

U.S. Environmental Protection Agency. "Climate Change." 2008. Retrieved September 20, 2008 (http://epa.gov/climatechange/basicinfo.html).

Index

A

allotropes, 8, 13
alloys, 13
American Dental Association, 36
atomic number, 20
atomic structure, 19–20, 22–23

B

Berzelius, Jakob, 10

C

carbon
  atomic structure, 19–20, 22, 23,
    24–25
  and climate change, 35
  compounds, 24–25, 27
  forms of, 8–9, 10, 33–34
  functions, 4–5, 8–9, 24, 32
  history of, 8, 17
  properties, 8–9, 11, 23
  symbol, 4
carbon cycle, 25, 35
carbon dating, 32–33
carbon group elements
  in everyday life, 32–37
  list of the five, 4, 6
  on the periodic table, 4, 6, 7, 22
  and tomorrow, 37
carbon monoxide, 33–34
chemical bonds, defined, 6, 23

chemical properties, defined, 6
compounds, defined, 6
covalent bonds, defined, 23

D

diamonds, 8–9, 10

E

electrodes, 15

F

fossil fuels, 9, 35, 37

G

galena, 15, 31
germanium
  atomic structure, 20, 22
  compounds, 30
  functions, 5, 12, 30
  history of, 12, 17
  properties, 11, 12
  symbol, 4
global warming, 35
graphite, 8–9

H

hydrocarbons, 25, 27

I

infrared detectors, 12
ionic bonds, defined, 23

K
Kipping, Frederic, 29

L
lead
  atomic structure, 20, 22
  compounds, 31
  forms of, 15
  functions, 5, 15, 31, 36–37
  history of, 13, 15, 17, 31
  properties, 10, 11, 15
  symbol, 4, 13, 15
lead poisoning, 15, 31, 37

M
malleable, defined, 13
Mendeleyev, Dmitry, 29
methane, 23, 27

O
organic chemistry, 24
oxides, 6

P
periodic table, 4, 6, 7, 16, 20, 22, 29
photosynthesis, 25
physical properties, defined, 7

R
respiration, 25
Rochow, Eugene, 29

S
semiconductors, 10, 12, 30
silicon
  atomic structure, 20, 22, 27, 29
  compounds, 27, 29
  forms of, 5, 9–10
  functions, 5, 10–11, 29, 33, 34–36
  history of, 10, 17, 29
  properties, 10, 11
  symbol, 4
solar cells, 11
solder, 13, 15

T
tin
  atomic structure, 20, 22
  compounds, 30–31
  forms of, 13
  functions, 5, 13, 30–31, 36
  history of, 13, 17
  properties, 10, 11, 13
  symbol, 4

U
ununquadium, 22
U.S. Environmental Protection Agency, 34
U.S. Food and Drug Administration, 36

V
valence electrons, 22

W
white tin, 13
Winkler, Clemens, 12

About the Author

Brian Belval has a bachelor's degree in biochemistry from the University of Illinois. He worked as a research scientist for a number of years before returning to school to study literature at the University of Massachusetts. He currently lives in New York City, where he combines his interest in science and writing as an editor of medical textbooks.

Photo Credits

Cover, pp. 1, 18, 21, 26, 38–39 Tahara Anderson; p. 7 © sciencephotos/ Alamy; p. 8 © Susumu Nishinaga/Photo Researchers, Inc.; p. 10 © Science Museum/SSPL/The Image Works; p. 12 © www.istockphoto.com/ Stephen Pottage; pp. 14, 28 Wikimedia Commons; p. 17 NASA; p. 19 © Matthias Kulka/zefa/Corbis; p. 30 © www.istockphoto.com/ Martina Misar; p. 33 © Philip Gould/Corbis; p. 34 © Paul A. Souders/Corbis; p. 36 © The Irish Image Collection/Corbis; p. 37 © www.istockphoto.com/Martina Tannery.

Designer: Tahara Anderson; Editor: Nicholas Croce;
Photo Researcher: Amy Feinberg